90 Seconds in the Space

90 Seconds in the Space

Yevtushenko Yevhenii

Library of Congress Control Number:		2014911803
ISBN:	Hardcover	978-1-4990-4472-0
	Softcover	978-1-4990-4473-7
	eBook	978-1-4990-4474-4

This book was printed in the United States of America.

Rev. date: 06/27/2014

To order additional copies of this book, contact:
Xlibris LLC
1-888-795-4274
www.Xlibris.com
Orders@Xlibris.com
636458

Tardigrade

Dioecious, incredible creatures they dare to be even more astonishing than their appearance. While in a state of cryptobiosis organisms are able to resist environmental extremes that would be instantly lethal to the animal if in the active state. As science obtains a better understanding of biological processes we must at times re-examine previous beliefs or understandings. This is perhaps exemplified by cryptobiosis. The issue pertains to the question of whether or not tardigrades can die and come back to life. The answer is no. However, Tardigrades In Space or "TARDIS" is the first research project to evaluate the ability of tardigrades to survive under open space conditions.

Introduction

Here I am, writing New York worst seller. It is hard to get an idea what's going to happen to a human in the space by dancing around this significant subject. All kids have already gone to bed, early spring, your cat is taking a short night walk and the story has begun. Nobody knows anything unless they say opposite. They don't even know how you could get there. Every given empty page is a great gift for us all. Each page for each second in the open space you have got left.

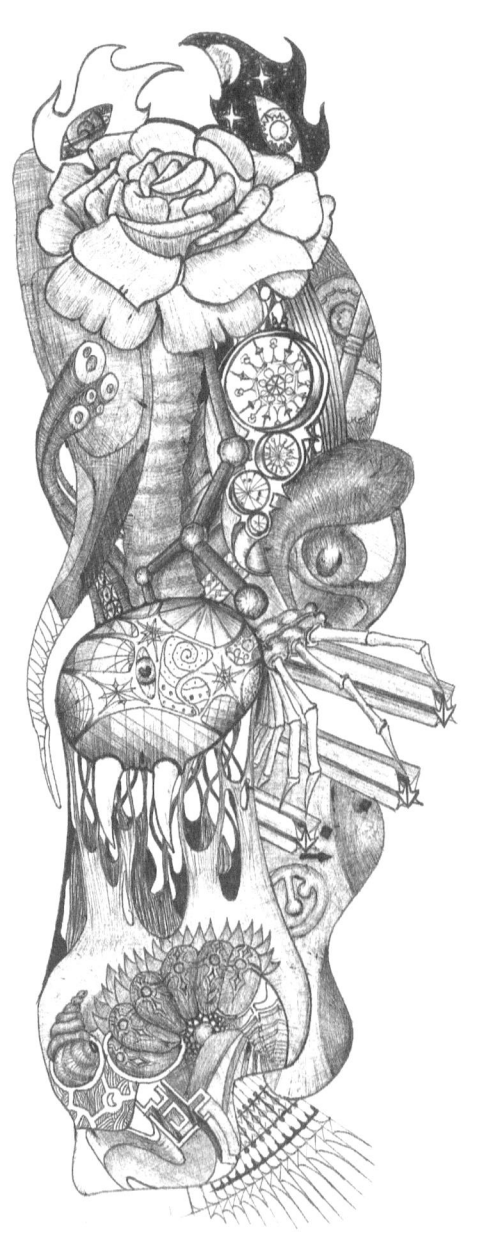

www.ingramcontent.com/pod-product-compliance
Lightning Source LLC
Chambersburg PA
CBHW030905180526
45163CB00004B/1715